GORILLAS

And More!

Bright Connections Media
A World Book Encyclopedia Company
233 North Michigan Avenue
Chicago, Illinois 60601
U.S.A.

For information about other BCM publications,
visit our website at http://www.brightconnectionsmedia.com
or call 1-800-967-5325.

Gorillas and More!
ISBN: 978-1-62267-001-7

Printed in China by Toppan Leefung Printing Ltd.,
Guangdong Province
1st printing July 2012

Cover photo: © Minden Pictures/Masterfile

© Jeroen Hendriks, Alamy Images 47; © Rob Henderson, Getty Images 49; © National Geographic/Getty Images 59; © iStockphoto 57; © Minden Pictures/Masterfile 13, 19, 21, 23, 25, 33, 41, 45; © Shutterstock 7, 17, 27, 37, 43, 45, 55, 61; © F1 Online/SuperStock 31; © Gerard Lacz Images/SuperStock 53; © Minden Pictures/SuperStock 39; © NHPA/SuperStock 35; © Tier und Naturfotografie/SuperStock 51; © Thinkstock 29

Maps, diagrams, and illustrations by WORLD BOOK

GORILLAS

And More!

By Karen Ingebretsen

A World Book Encyclopedia Company

CONTENTS

As you read, you may come across words you don't know. You can find the definitions of many words from this book in the Gorilla Talk section on page 63.

WHAT IS AN APE?

An ape is a member of the group of animals called primates. The primate group also includes monkeys and human beings.

There are five main kinds of apes—bonobos *(buh NOH bohz)*, chimpanzees, gibbons, gorillas, and orangutans. All these animals have hairy, tailless bodies, and their arms are longer than their legs. Apes also have long fingers and toes. They have large brains and are very intelligent, too.

Many people confuse apes and monkeys, but the two groups differ in several ways. Most monkeys have tails, but apes do not. Monkeys are usually smaller than apes and have shorter fingers and toes. Monkeys also seem less intelligent than apes.

Apes resemble humans in body structure more than any other animals do. Most scientists believe that apes are more closely related to human beings than they are to monkeys.

Silverback (adult male)
lowland gorilla

WHERE IN THE WORLD DO GORILLAS AND OTHER APES LIVE?

Gorillas live in the tropical lowland forests of western and central Africa. They also live in the mountain forests of eastern Africa.

Some other apes live in Africa, as well. Bonobos live in a section of African rain forest south of the Congo River in the Democratic Republic of the Congo (DRC). Chimpanzees live in a wide range of habitats from western to eastern Africa.

Apes also live in Asia. Orangutans live in the tropical forests on the islands of Borneo and Sumatra in Southeast Asia. Gibbons live in the forests of northeastern India and in Southeast Asia.

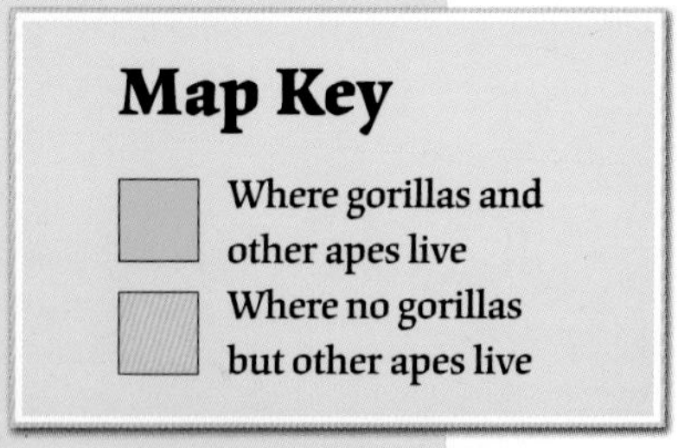

Map Key

Where gorillas and other apes live

Where no gorillas but other apes live

WORLD MAP
N
W
E
S
Arctic Ocean
North America
Atlantic Ocean
Europe
Asia
Pacific Ocean
Africa
Equator
South America
Indian Ocean
Australia
Pacific Ocean
Antarctica

HOW DO APES MEASURE UP?

Gorillas are the largest members of the ape family. A large male gorilla can weigh 390 pounds (177 kilograms). Standing up on its legs, it might be 6 feet (1.8 meters) tall. Females weigh about 200 pounds (91 kilograms) and are shorter than males. In contrast to gorillas, gibbons—the smallest members of the ape family—each weigh about 15 pounds (7 kilograms) and are about 3 feet (0.9 meter) tall.

A male orangutan weighs about 180 pounds (82 kilograms) and stands about $4\frac{1}{2}$ feet (1.4 meters) tall. Females are about half as large.

A male chimpanzee weighs about 110 pounds (50 kilograms), and a female weighs about 90 pounds (41 kilograms). Chimpanzees range in height from $3\frac{1}{4}$ feet (1 meter) to $5\frac{1}{2}$ feet (1.7 meters). Although they are about the same height as chimpanzees, male and female bonobos weigh about 75 to 100 pounds (34 to 45 kilograms).

TYPES OF APES

gibbon

orangutan

gorilla

bonobo

chimpanzee

WHAT'S SO "GREAT" ABOUT THE GREAT APES?

Primarily, it's their size. "Great apes" include bonobos, chimpanzees, gorillas, and orangutans. Gibbons, which are much smaller than the others, are called "lesser apes."

In addition to their size, there are other special things about great apes. They rank as the most intelligent animals after human beings. Some apes—chimpanzees in particular—use simple tools in some of the same ways humans do. For example, chimpanzees use stones to crack open the shells of hard nuts. They can then eat the nuts that were held inside. In addition, laboratory researchers have been able to teach chimps to perform tasks and to communicate with humans.

Chimpanzee using a rock to crack open nuts

WHAT'S UNDER ALL THAT HAIR?

Apes have strong bones and powerful muscles. Their body structure is more like a human's than that of any other animal. Apes and humans have similar bones, muscles, and organs. But apes walk on four limbs and have shorter legs and more body hair than humans.

Both humans and apes have highly developed nervous systems and large brains. Like humans, apes rely on their excellent eyesight for much of their information about their environment. They have large eyes and stereoscopic vision. This means that they have the ability to see depth.

Like humans, apes have opposable thumbs. This means that their thumbs can be placed opposite their fingers so that they can grasp objects. Apes also have nails instead of claws on their fingers and toes.

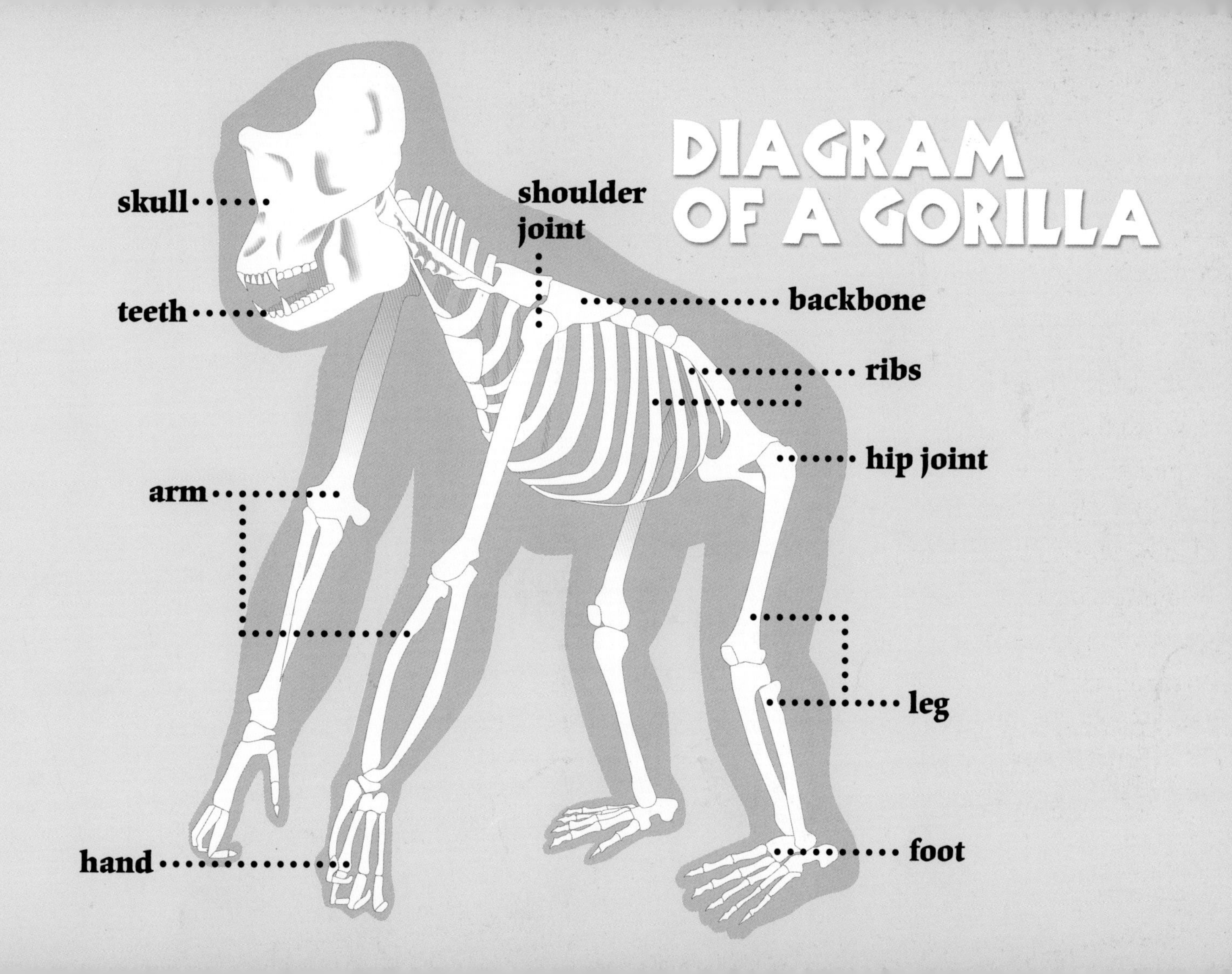
DIAGRAM OF A GORILLA
skull
teeth
shoulder joint
backbone
ribs
hip joint
arm
leg
hand
foot

ARE GORILLAS THE "GREATEST" OF THE GREAT APES?

In terms of size, yes. Gorillas are the largest of the great apes. Gorillas have huge shoulders—nearly twice as wide as those of chimpanzees. Gorillas also have a broad chest, long arms, and short legs.

Today, most scientists agree that there are two *species* (kinds) of gorilla. They are the eastern gorilla and the western gorilla. As their names suggest, each lives in a different area of Africa. However, not all gorillas look alike. Some differences include color and length of hair, size of jaw, shape of face, and length of arms.

Because of these differences, scientists have divided gorillas into four varieties, or subspecies. Two varieties of eastern gorilla are the eastern lowland gorilla and the mountain gorilla. Two varieties of western gorilla are the western lowland gorilla and the Cross River gorilla.

Male lowland gorilla

ARE GORILLAS AS FIERCE AS THEY LOOK?

When people think of a gorilla, they might think of a ferocious beast, bearing its teeth and beating its chest with its fists. Male gorillas sometimes behave this way when they want to show their strength. But a display like this actually helps them end or even avoid a conflict with another gorilla. Have you heard the expression, "the best defense is a good offense"? Gorillas usually follow this advice. The gorilla that looks stronger often drives the intruder away without a fight.

Male gorillas do sometimes fight over territory or mates. But gorillas are mainly gentle animals that need companionship and attention. They are actually rather shy. A gorilla will not hurt a human being unless it is threatened or attacked.

A silverback mountain gorilla displays his teeth to frighten a rival.

HOW DO GORILLAS COMMUNICATE?

Gorillas use a variety of sounds to communicate. They laugh; they grumble softly to show they are contented or in agreement; they whimper or cry when saddened; they make a barking noise when they are annoyed; and they scream when angered.

Gorillas can hoot loudly enough to be heard at least ½ mile (0.8 kilometer) away. Gorilla males hoot to let other troops in the area know where they are. It sends the message to keep away!

Beating their chest is another way that gorillas communicate. They do this to show off their strength to others who might challenge them. Most of the time this display is just a bluff!

Young mountain gorilla
chest-beating display

WHAT DOES A GORILLA DO ALL DAY?

A gorilla's typical day is what some people would call ideal. A gorilla has a morning meal, takes a nap, and travels in the afternoon. Most of a gorilla's daylight hours are spent feeding or resting. As sunset approaches, a gorilla prepares for bedtime by making a nest.

Some gorillas build their nests on the ground. Others build them in trees. A gorilla's nest is simple. It is made up of whatever leaves, grasses, and branches are nearby. A nest can be built in as little as two or three minutes. Each nest is used only one time, because gorillas sleep in a different place every night. They are always on the move.

Adult gorillas climb trees only to reach fruit or leaves, to get a better view of something, or to sleep at night. Young gorillas climb more often than adults do. They play by climbing or swinging on branches and vines. They wrestle with one another, too.

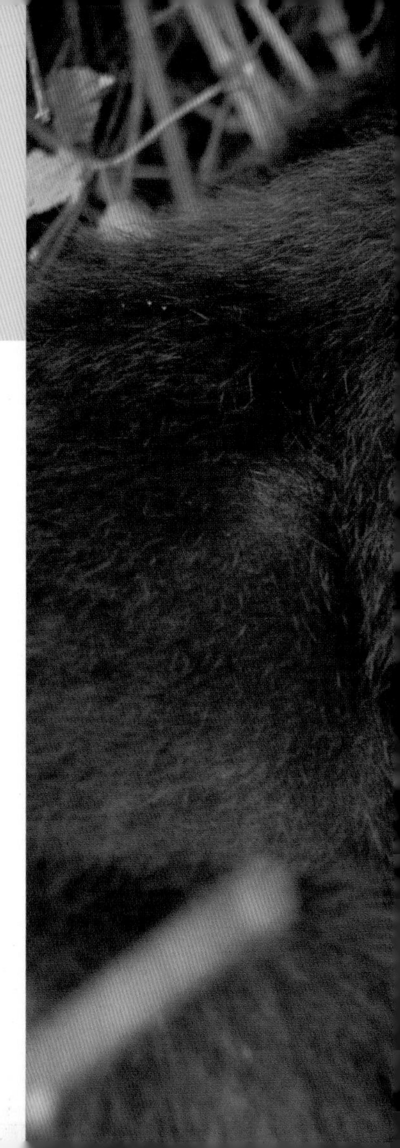

Mountain gorillas at rest

WHAT IS FAMILY LIFE LIKE FOR A GORILLA?

Gorillas travel through the forest in groups called troops. A troop can have anywhere between 2 and 34 members. Troops may have one or more adult males, two or more adult females, and several young gorillas. An adult male always leads the troop. He makes all the decisions, such as when to get up in the morning, where to go, and when to rest. This male also protects the troop against danger. The fur on the back of this adult male is gray or silver. He is called a silverback.

The other males in a troop sometimes leave and wander alone. A new or enlarged troop forms when one or more females leave the troop of their birth to join another troop or a lone male.

Each troop wanders around in its own home range, which is from 2 to 15 square miles (5 to 39 square kilometers) in area. Several troops may live in the same area of a forest, but they usually avoid one another.

Mountain gorilla troop

WHAT'S FOR DINNER?

A gorilla's diet is almost all vegetables. You probably wouldn't want to share a gorilla's dinner. Unless, of course, you like eating leaves, bark, shoots, stems, roots, flowers, and fruit.

The choice of foods varies, depending on the gorillas' habitat. Mountain gorillas eat more leafy plants and herbs that they find near the ground, and lowland gorillas climb to find ripe fruit.

Most of the regular plant life in a gorilla's habitat is a poor source of nutrients and calories, which the animal needs to grow and stay healthy. For this reason, a gorilla will even eat small amounts of soil if it contains needed minerals.

Female western lowland gorilla eating bark

WHAT ARE INFANT GORILLAS LIKE?

When they are born, infant gorillas are at about the same level of physical development as human newborns. They are weak and uncoordinated, and their senses are not sharp. However, gorillas develop much more rapidly than humans do. Baby gorillas can sit up and crawl sooner than human infants can. By 5 months of age, gorillas can walk.

Very small infant gorillas are carried near their mother's chest. As it grows and becomes heavier, a baby spends more time riding "piggyback" on the adult's back.

Gorilla births do not occur very frequently in any family group. So when an infant is born, both male and female adult gorillas of all ages are very interested in the new arrival. When the group is resting, the young continue to play and may climb over sleeping adults. The adults do not seem to mind this at all and let them play.

Western lowland gorilla
mother and infant

DO OTHER APES LIVE AND TRAVEL IN GROUPS?

Chimpanzees also form loosely tied groups called communities. Members of a community share the same territory, or home range. Within these communities, the apes travel alone or in smaller groups that vary in number and change members often. Males and females within a group have almost complete freedom to come and go as they wish. Each chimpanzee has its own favorite companions.

There are three types of groups: (1) all male bands, (2) bands of mothers and their infants, and (3) mixed bands of both male and female chimpanzees.

Bonobos, which are closely related to chimpanzees, also form communities. Within a bonobo community there are social groups of 7 to 10 members (both males and females).

A mixed band of chimpanzees on the move

HOW DO CHIMPANZEES USE TOOLS?

Chimpanzees make and use simple tools more than any other creatures except human beings. For example, chimpanzees strip the leaves from plant stems and use the stems as tools to catch termites to eat. They also use leaves as "sponges" to soak up water to drink, and as "tissues" to clean their bodies. Some chimpanzees use stones as "hammers" to crack open nuts.

Young chimpanzees use sticks or stones to tickle themselves in hard-to-reach places. Chimpanzees wave sticks in the air to frighten or ward off enemies. Sticks are also used as hooks to pull down branches so the animals can reach the fruit on them.

Chimpanzees practice making and using tools from the time they are very young. They imitate the actions of adults, getting better over time. What begins as simple play becomes an important skill in later life.

Young chimpanzee catching termites with a stem

HOW DO CHIMPANZEES COMMUNICATE?

Chimpanzees communicate using barks, grunts, hoots, and screams. If they find a large food supply, the apes jump through the trees, hoot loudly, and beat on tree trunks. This lets all other chimpanzees within hearing distance know there is food available.

Chimpanzees also communicate with body postures, facial expressions, and hand gestures. Chimpanzees greet each other by embracing or by touching various parts of the other's body. Their facial expressions show many emotions, including excitement, fear, and rage. A chimpanzee, especially a male, will sometimes scream and wildly run around to intimidate another chimpanzee. This noisy show is known as a "display."

Scientists have worked to teach chimpanzees sign language and other types of language. In Atlanta, a chimpanzee named Lana learned to use symbols on a computer keyboard to ask for food, company, and music.

Male chimpanzee in a charge display

HOW ARE CHIMPANZEES AND HUMANS ALIKE?

Scientific evidence suggests that chimpanzees (and their close cousins, the bonobos) are more closely related to humans than any other animal. Not only are they our nearest relative, we are theirs. We are a closer relative of the chimpanzee than is the gorilla, orangutan, or any other kind of primate.

Scientists believe this because human and chimpanzee DNA are nearly identical. DNA is a molecule that determines hereditary traits and affects how organisms look and behave. DNA is found in the cells of all animals. Closely related animals share identical DNA.

Even without examining their DNA, it is easy to see that chimpanzees are closely related to humans. Both have similar body structures. Both also have many complicated behaviors, such as using tools and communicating their emotions.

Human and ape hands

WHY DO CHIMPANZEES GROOM EACH OTHER?

Adult chimpanzees spend a part of each day in a friendly social activity called grooming. During this time, two or more chimpanzees sit and pick through each other's hair. They remove dirt, insects, leaves, and burs (the seeds of certain plants) from each other.

In addition to helping keep each other clean, grooming reduces tension among group members. Chimpanzees are sociable animals who seem to need physical contact, and grooming helps satisfy this need. They enjoy it, too. Grooming also strengthens ties between group members. Sometimes as many as 10 chimpanzees will take part in these peaceful, relaxed grooming sessions. Often a grooming group will be made up of a mother and several of her offspring of different ages. Adult males may also groom one another.

Chimpanzees grooming

HOW ARE BONOBOS AND CHIMPANZEES RELATED?

Scientists once considered the bonobo to be a smaller variety of the chimpanzee. In fact, bonobos were known as pygmy chimpanzees. Today, however, the bonobo and chimpanzee are considered separate species of great apes.

At first glance, bonobos look much like chimpanzees, but there are some differences. Bonobos are more slender and have longer legs. They have smaller, rounder heads, and their faces are blacker than those of chimpanzees.

Bonobos, like chimpanzees, form communities of males and females. But, whereas the males lead chimpanzee communities, bonobo societies center around strong alliances of females.

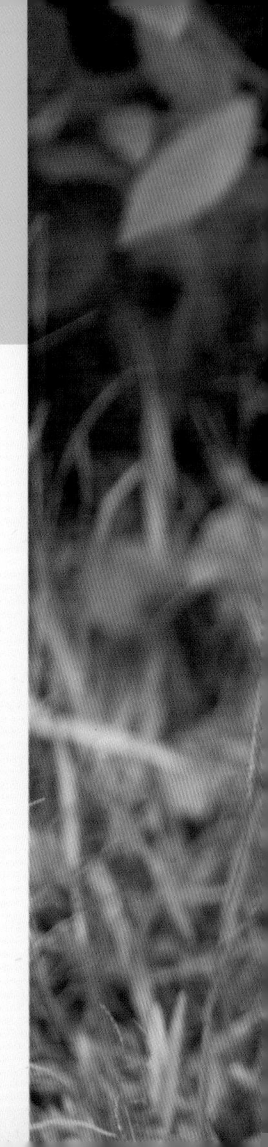

Young male bonobo

WHICH APES ARE LONERS?

Orangutans are loners. Unlike other great apes, they spend most of their time by themselves.

Scientists think that orangutans spend most of their time alone because of their feeding needs. Orangutans eat mainly fruit. Often, too little food is available at any one spot to feed a whole group or to feed even a single orangutan for very long. Therefore, an orangutan must be always on the move, looking for the next source of ripened fruit for itself and any youngsters it has. However, at times when ripe fruit is plentiful, several orangutans have been seen eating peacefully in the same tree.

Sometimes, orangutans will meet each other by chance as they wander through the treetops. When two males meet, the stronger will drive the other away. When females meet, the encounter is nearly always peaceful.

Female orangutan

WHERE DO ORANGUTANS SPEND MOST OF THEIR TIME?

Unlike gorillas, chimpanzees, and bonobos, orangutans spend most of their day high up in the trees. Orangutans are the largest tree-living animals in the world.

As they travel high above the forest floor, orangutans make little noise. It is hard to spot and study them because they move so silently and are usually high above the ground. Their reddish-brown hair blends in with brown tree trunks and branches.

Orangutans move awkwardly when they are on the forest floor. When walking on all fours, chimpanzees and gorillas use their knuckles for support. Orangutans, however, support their upper bodies with closed fists. Sometimes orangutans will walk upright, too. And they tend to walk on the outside edges of their feet, not flat on the soles, because they have long, curved toes.

Female orangutan in
a forest canopy

DO MALE AND FEMALE ORANGUTANS LOOK ALIKE?

An adult male orangutan is twice the size of an adult female orangutan. The face of a male orangutan is also different. A mature male orangutan has broad cheek pads that jut out from the sides of his face. He also has a large throat pouch. A female has no cheek pads and only a small throat pouch.

Researchers aren't quite sure what purpose the cheek pads serve. But the throat pouch amplifies an orangutan's voice, which is useful when the male makes a "long call"—a series of roars to remind others that he owns a territory. Long calls can be heard more than $\frac{1}{2}$ mile (0.8 kilometer) away.

Both male and female orangutans have red hair. Because of their red hair, orangutans look very different from the other apes.

Female (left) and male (right) orangutans

HOW IS AN ORANGUTAN'S BODY SUITED FOR LIVING IN TREES?

An orangutan has extremely long arms, compared with its legs and upper body. Its arms reach to its ankles when it is standing upright. It has long, curved fingers and toes that help it grasp branches, and its shoulder joints and hip joints are remarkably flexible.

Orangutans can walk on the ground, but they rarely do so. When they do, they walk on all four limbs, slowly and cautiously, and they usually only walk as far as the distance from one tree to the next. They travel much more gracefully when swinging from branch to branch in the trees.

Because they live so dangerously high in the trees, a mother orangutan must be careful with her offspring. At first, an infant simply holds on as tight as it can to its mother. Soon, however, the mother encourages her young to climb. She usually stays within arms' reach until the young orangutan builds its strength.

Orangutan with infant

WHAT IS THE PURPOSE OF GIBBONS' "SONGS"?

Gibbons are well known for their ability to "sing." Songs vary from species to species, and males and females of the same species have different songs. By listening for differences in the songs, researchers are able to tell from a distance what kind of gibbon is singing.

Calls are usually sung as a "duet" by an adult male and adult female, with their offspring sometimes joining in. The female's part is usually the longest. It may be made up of whoops, booms, barks, or high-pitched calls. From start to end, the song may last for several minutes. The songs are meant to help claim a pair's territory.

Since they are about the same size, it is sometimes hard to tell a male gibbon apart from a female. Among some species of gibbon, the male and female each have a different color of fur.

White-handed gibbon singing

HOW DO GIBBONS GET AROUND?

Gibbons live high in the trees and rarely come to the ground. They do most of their traveling through the trees by swinging from branch to branch. They often use their hands like hooks, rather than using them to grasp the tree limb.

There are many species of gibbon. Some are called different names like Hoolock or Siamang.

Gibbons are the star acrobats of the apes. They often make swings as long as 10 feet (3 meters). A human child traveling across a set of horizontal monkey bars on the playground moves in a similar way, but for much shorter distances.

Gibbons also sometimes walk on top of tree branches using only their legs, similar to the way human beings walk on the ground.

White-handed gibbons swinging on a vine

DO GIBBONS LIVE IN FAMILY GROUPS?

Gibbons live in family groups that usually consist of a male, a female, and up to four of their young. The offspring stay with their parents until they are about 5 to 6 years old. At that time, they begin to look for mates and territories of their own. They may spend years searching for a mate.

A new gibbon may be born to parents about every 2 to 3 years. Only one baby is born at a time. For the first year of its life, the mother guards a baby closely. In some species, the mother continues to be the main caretaker in the young gibbon's second year. However, in other species, the father will help out.

More so than any other apes, male and female gibbons that have mated with each other remain as a couple, caring for their offspring. Gibbons can live for up to 40 years in the wild.

White-cheeked gibbon family. Among these gibbons, females are light-colored while the males are black.

DOES EACH GIBBON FAMILY HAVE ITS OWN TERRITORY?

A gibbon family has a small territory which it defends. But gibbons rarely use physical contact to do so. Instead, they use calls or songs to warn other gibbons to stay away.

A family's territory must be large enough to provide "necessities." Necessities include separate sleeping trees for all family members and plenty of food, even during times when food is most scarce.

Territories for these apes vary in size, depending on species and location. The smallest ranges are about 25 acres (10 hectares) and the largest are about 130 acres (53 hectares).

Gibbon family in
a forest canopy

HOW HAVE PEOPLE LEARNED ABOUT APES?

Scientists study apes both in captivity and in the wild. Some researchers spend many years of their lives living among wild apes in their native habitat and carefully watching their behavior.

Louis Leakey (1903 - 1972) was a British anthropologist (a scientist who studies humanity and human culture) who searched in Africa for evidence of the earliest human beings. He also helped three women—Jane Goodall, Dian Fossey, and Biruté Galdikas—begin detailed research of great apes living in their natural habitats.

Jane Goodall (1934 -), a British zoologist, began working with chimpanzees in the wild in 1960. Most of her study has taken place in northwestern Tanzania. Dian Fossey (1932 - 1985) was an American zoologist who began studying mountain gorillas in east-central Africa in 1966. Biruté Galdikas (1946 -), a Lithuanian-born Canadian anthropologist, began studying orangutans in Borneo in 1971.

Jane Goodall studying chimpanzees in Tanzania, 2006

ARE APES IN DANGER?

Most kinds of apes are in danger of extinction (the dying off of all their kind). Today, there are believed to be about 400,000 great apes in Africa and Asia, compared with more than a million in the 1800's. Without help, some species of apes may die off completely in our lifetime—in 50 or fewer years.

Sadly, it is the apes' closest relatives—human beings—who pose the greatest threat to their survival. Humans destroy the habitat of these creatures by clearing forests to build homes, establish farms, and sell wood. People capture these animals for sale to zoos and research centers, to be kept as pets, and for the sale of parts of their bodies. Apes are also hunted for food.

But humans also offer the greatest hope for saving the apes. The United Nations—an organization of nations working for world peace and security—has created a plan to help protect the apes. The United Nations hopes to educate people about the risks of ape extinction and the benefits of preserving ape habitats.

Female mountain gorilla

APE FUN FACTS

- Until the mid-1800's, Europeans didn't know gorillas existed.
- Scientists do not think gorillas can swim.
- In 1961, a chimpanzee named Ham rode aboard a Mercury space capsule during an 18-minute test flight. He returned safely to Earth.
- Chimpanzees mainly eat fruits and leaves, but they will also hunt and eat small animals, including monkeys.
- Human and chimpanzee DNA is between 95 and 99 percent identical.
- The way gorillas, chimpanzees, and bonobos walk is called knuckle walking.

ape A chimpanzee, bonobo, gorilla, or orangutan (great apes), or a gibbon (lesser ape).

bonobo An African ape closely related to the chimpanzee.

calorie A unit used to measure the amount of energy in food.

captivity The condition of being held in a place removed from a natural habitat, such as in a zoo.

community A loosely tied collection of groups of chimpanzees or other apes.

DNA The molecule inside the cells of animals and plants that carries information on traits inherited from the parents.

extinction The dying off of an entire group of animals or plants.

grooming A social behavior in which primates remove dirt, insects, or other material from each other's hair.

habitat The area where an animal lives, such as a grassland or forest.

heredity The passing of biological traits from parents to offspring.

long call A series of roars made by a male orangutan to announce his claim to a territory.

lowland forest Forested land that is lower and flatter than the surrounding land.

opposable thumb A thumb that is opposite to the rest of the fingers, allowing the hand to grasp objects.

primates The group of animals that includes monkeys, apes, and human beings.

species A group of the same kind of animals.

stereoscopic vision The ability to see depth.

territory An area in which an animal lives and from which it keeps out many other animals.

troop A group of gorillas or certain other primates.

INDEX